Published by Smart Apple Media
1980 Lookout Drive, North Mankato, MN 56003

Designed by Stephanie Blumenthal
Production Design by Kathy Petelinsek

Photographs by Maslowski Wildlife Photography

Library of Congress Cataloging-in-Publication Data

Maslowski, Stephen. Birds in fall / by Steve Maslowski with Adele Richardson.
p. cm. — (Through the seasons ; 3)
Summary: Focuses on the migration process of birds in the fall including descriptions of
what they need in order to migrate, how they prepare for the trip, and how they travel.
ISBN 1-58340-058-3
1. Birds—Migration—Juvenile literature. 2. Birds—Juvenile literature.
[1. Birds—Migration.] I. Richardson, Adele. II. Title.

QL698.9 .M38   2001
598.156—dc21   99-046947

First Edition

2 4 6 8 9 7 5 3 1

THROUGH THE SEASONS

BIRDS IN FALL

Text by Steve Maslowski with Adele Richardson
Photographs by Maslowski Wildlife Photography

SMART APPLE MEDIA

The season of fall is also called autumn. It is a very colorful time of year. Leaves on many trees change from green to different shades of gold, orange, red, and brown. Soon after changing color, the leaves drop to the ground, and the trees are left dull and bare. Fall starts on the 22nd day of September, bringing cooler temperatures and shorter days.

Autumn is also called "fall" because it is the time of year that leaves fall from trees.

The skies over North America are very active during this season, as many different types of birds fly to the warm southern part of the continent. These long flights are called migrations. Many birds migrate to avoid the snow and cold temperatures of the coming winter season in the North. When spring arrives, they will make the long trip back home. In most

Robin drinking

cases, only those birds that live in northern regions fly south. If a bird already lives in a warm area, it has no need to migrate.

Birds are one of the few animals that can fly, and they are perfectly built for it. All of a bird's feathers point toward its back. This gives the bird a streamlined body, allowing it to easily cut through the air without the wind ruffling its feathers the wrong way.

5

Temperatures can drop early in the fall. On September 24, 1926, the temperature dipped to –9° F (–23° C) in Montana.

Flocks of ducks (top) and geese (bottom)

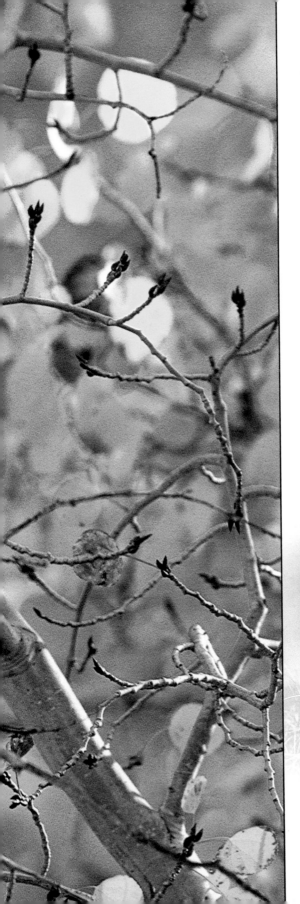

The bones of a bird's skeleton are not as heavy or solid as those of most animals. Instead, the bones are hollow and lightweight. Many of the bones are also fused, or joined together; this makes the skeleton very strong. And unlike many animals, birds do not have heavy jawbones and teeth. Rather, a bird has a beak that is light and slender.

Every year, the golden plover migrates very long distances. The bird starts its trip at the northernmost tip of North America, then flies every day until it reaches the southernmost part of South America. In the spring, the plover makes the long flight back to its home. In one year, it may fly more than 20,000 miles (32,200 km)!

Steller's jay

Flying takes a lot of muscle. To flap its wings, a bird relies on powerful chest muscles. The muscles attach the wings to a ridge on the breastbone called a keel. These big muscles may make up as much as one-third of a bird's total weight.

wing. The top side of the wing is slightly rounded, while the underside is flat. When the bird is flying, air flows faster over the curved top. This lowers the air pressure on the top side of the wing. It also creates more air pressure under the wing, lifting the bird upward. The wings of an airplane work in the same way.

Perhaps the most important key to flying is the shape of the bird's

To get ready for their migration, birds eat a lot of food in the fall.

Great gray owl in flight

Fruits, berries, and nuts of all kinds are ripe and ready to be eaten. Wildflowers that have bloomed all summer also produce seeds during this season. Birds gobble down all the food they can find. The extra weight they gain during this time helps them to store up energy for the long flight ahead.

The ruby-throated hummingbird often crosses the Gulf of Mexico during its migration. This means that the tiny bird must fly about 600 miles (970 km) without stopping. It may take the hummingbird up to 24 hours to make the long crossing, and the bird has nowhere to land if it becomes tired.

While birds gain extra weight for their migration, they also replace their feathers. During the summer, it is common for feathers to become dirty or damaged. In the fall, birds will lose these feathers. Then, strong new feathers will grow in to take their place. This changing of feathers is called molting.

The goldfinch is a small bird that changes color during the fall. In the summer, the male is a bright yellow, and the female is an olive green and dull yellow. When the goldfinches molt in the fall, their new feathers grow in a dull brown color. This helps them to hide from enemies.

Blue jay carrying an acorn

Birds do not lose all of their feathers at one time. Instead, molting takes place over the course of several weeks so that no bare spots appear on a bird's body. Birds usually start to molt near the end of summer and have new feathers ready in time for their migration.

The Canada goose is a large bird that can weigh up to 20 pounds (9 kg). It has a gray-brown body and a black neck and head. Geese spend the summer in the northern parts of North America. In the fall, flocks of geese gather together in V-formations to migrate south, sometimes flying as far as Mexico.

Canada geese in cornfield

Common loon

In some types of birds, the new feathers look exactly like the old ones. Other birds completely change color. Loons and warblers, for example, are brighter and more colorful during the summer months. The new feathers they grow in the fall are very plain. This new, dull coloring helps the birds hide from enemies.

Once birds have fattened up and grown their new feathers, they are ready to migrate. Many scientists think that birds rely on signs in nature—such as dropping temperatures and decreasing daylight hours—to tell them when it's the right time to leave.

In the northernmost part of North America, smaller birds that live along the seashore are among the first to migrate. They start their trip in small groups of just a few families. More and more birds join the groups along the way until huge flocks fill the sky.

Shorebirds before migration

Surprisingly, most smaller birds make their journey at night. They stop to rest and find food during the day. Scientists are not certain why small birds do this. Perhaps it is because most of their enemies sleep at night, which makes travel during that time safer. Some of these enemies are large birds—including

The red-throated loon can grow up to 25 inches (64 cm) long. It lives in Canada during the warmer months of the year. The loon visits the United States only during the winter, when it lives along the Pacific or Atlantic coasts. The red-throated loon earned its name from the chestnut-colored feathers on its neck.

Flock of snow geese

The whooping crane is one of the tallest birds in North America, often growing more than four feet (1.2 m) tall. It has a white body, black-tipped wings, and a red head and face. Most whooping cranes spend the spring and summer in Alberta, Canada, and then migrate to Texas for the winter.

some kinds of hawks and eagles— that fly during the day as they make their way south as well. An average small bird can travel up to 200 miles (322 km) in one night.

Many birds migrate without being seen, and it's not just because they are traveling at night. Birds usually stay very high in the sky while they migrate, soaring at a height of any- where from 3,000 to 6,000 feet

(914–1,829 m) above the ground. Some birds, such as geese, have been known to fly at a height of more than 20,000 feet (6,096 m)!

~~~

How birds navigate, or find their way, is not fully understood. Somehow, they always seem to know exactly where they are going. Scientists have come up with several theories to explain this mystery.

An Indian summer is hot, summer-like weather that occurs during the fall season.

Birds that travel by day may follow landmarks they remember from previous trips. These landmarks could be rivers, coastlines, or mountains. Other birds may use the sun to help guide them. Young birds that are migrating for the first time may simply follow older birds in the flock.

~~~

Those birds that fly at night may rely on the stars as a guide. Some scientists also believe that birds have a built-in "compass" that lets them know which direction they are heading at all times. If this is true, then birds may have more than one way to navigate in case clouds cover up the sun or stars.

Gray-cheeked thrush

However birds find their way during migrations, most scientists agree that they do it instinctively. This means that birds don't have to learn how to find their way during migrations—they have that ability from the moment they are hatched.

For every 1,000 feet (305 m) a bird flies up into the sky, the temperature drops 3.6° F (2° C).

The season of fall lasts until the 21st day of December. By then, most birds that migrate will have already finished their trip. They will spend the winter in an area where temperatures are warmer and food is more abundant. They will not have to make another long trip for a few months.

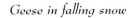
Geese in falling snow

INDEX